Some of the other books by Press Nip Impressions

On Employment *(2004)*

The Osage Mill *(2005)*

The Lazy Project Engineer's Path to Excellence *(2006)*

Raising EBITDA: the lessons of Nip Impressions *(2014)*

Advertising Arguments 2016 *(2015)*

Personalities in the Pulp & Paper Industry *(2017)*

Paperitalo's Pretty Good Pulp & Paper Mill Directory *(2017)*

The Proper Production (or Project) Meeting *(2018)*

Pulp Rats: An Unpleasant Tale *(2019)*

A Monograph on Purchasing *(2020)*

(indicates first printing)

I hope you enjoy this book

This book is dedicated to

Kevin Fedor,

the investment banker

that supported my early efforts

in this field

Contents

Introduction

This book is intended as a primer for the investor or investment banker not familiar with the pulp and paper industry. Your interests may lie in debt, equity, or both. This book will not focus so much on the types of investment instruments in which you are interested, but rather on the workings of the pulp and paper industry and how these manifests themselves in the public and private debt and equity markets. The author feigns no special expertise in financial instruments, although his experiences over the years have exposed him to the uses and ramifications of many. Rather, in this book he strives to give you a brief education in the investment capital, operating capital, and cash flow characteristics of this industry. Along the way, it is hoped you will also pick up some signals that will give you first clues as to whether a particular investment opportunity holds promise or is something which may be fraught with danger.

The author has worked in the pulp and paper industry for over fifty years as of this writing. Starting out as an engineer with the Procter & Gamble Company, his stints at Westvaco, Packaging Corporation of America, Jaakko Pöyry, and since 1992, as a private consultant to the investment community, have provided exposure, experience, and insight in nearly all grades and on all continents. He has worked for private and public companies and financiers, on equity and debt offerings, in rated and unrated conditions. His testimony has been important for clients prevailing in multimillion dollar lawsuits. He has consulted with the US Senate Anti-Trust Committee. Starting in the early 00's, he has developed a suite of digital publications pertaining to the pulp and paper industry widely read throughout the world (see www.paperitalo.com). His valuation work has been instrumental in many successful investment and divestment decisions.

The pulp and paper industry is one of the oldest and earliest members of the industrial age. Hence, it possesses a long history, some of the ancient which still influences it to this day. It is also an industry of traditionally high factory wages, and thus attracts certain elements wishing to exploit this feature. Further, the industry uses natural resources loved by many—forests—and has, in the past, produced considerable amounts of pollution, both of these features drawing the attention of environmentalists. Also, the products of the industry are essential and ubiquitous, produced to similar standards world wide, and traded freely among many nations, creating unique marketing conditions.

Welcome to the fascinating and challenging world of pulp and paper.

Chapter 1
Industries of similar nature—Rock, Paper, Scissors

From an investment viewpoint, three industries are similar in attributes. These are

> Basic Cement Manufacturing (Rock),
> Steel (Scissors), and
> Pulp and Paper (Paper).

All three of these industries require heavy capital investment, significant operating capital, and a steady supply of raw materials. Basic cement is slightly different from the other two in that the manufacturing assets absolutely must be located directly adjacent to the raw material—a limestone quarry—due to transportation considerations. The other two have a bit more flexibility in location but are not exactly free to be located anywhere.

Some may think that chemical processing or plastics belong in the same category, but this is an erroneous assumption. The distinguishing differences are that chemical processing and plastics plants can be built to smaller initial and incremental scales. This difference tends to smooth capacity side changes and align them more closely to demand side realities in the chemical and plastics industries (see Chapter 13 for more details). Also, the products of the chemical processing and plastics industries are often liquids, or neoliquids, such as powders or granules, which can be handled in much different ways than the products of the steel and pulp and paper industries (The author concedes the products of the basic cement industry are also granular or powder in form, but other conditions override this attribute in the case of cement).

The rock, paper and scissors industries' manufacturing plants are some of the largest in the world. They are continuous processes covering large amounts of real estate. It takes several days to start them in operation and to shut them down, hence they run continuously with only occasional maintenance shutdowns. These processes are so massive they require their own utility plants, water intake and waste treatment plants, and maintenance shops, with maintenance stores far exceeding the scope and value of the local hardware store. In effect, they are small cities unto themselves.

In this world, there are obviously many opportunities to deploy capital intelligently, and conversely, stupidly. Successful operations deploy capital simultaneously intelligently and frugally, creating an operation that is sustainable for many years. However, this is not the whole picture, for in the balance of this book we will focus on the pulp and paper industry and discuss both "inside the fence" and "outside the fence" those conditions and attributes that define and sustain a successful player.

Chapter 2
The four components of the pulp and paper industry

When we speak of the pulp and paper industry, there are four basic components in which we are interested. These are

1. Raw materials
2. Pulp manufacturing
3. Paper manufacturing and
4. Converting

There is another entire collection of businesses closely related to this industry, which are often called allied industries. These are companies that supply other raw materials (besides timber and waste paper) and manufacture the equipment used in the pulp and paper industry. These allied industries are beyond the scope of this book, except as mentioned occasionally in passing.

Then, in another category there are the solid and engineered wood industries. Often closely related to pulp and paper due to raw materials, they operate in a much different manner and are not within the scope of this book. This does not mean one will not occasionally find manufacturing facilities related to these industries adjacent to or on a pulp or paper mill site. They are there due to the symbiosis of raw materials. Their operational and economic behavior, however, can be quite different than those of the subject here.

It is worth taking a moment to look at the overall picture of the pulp and paper industry on a qualitative basis as relates to capital investment and potential for operating profits. In this simple table, more $'s means more money—a bad thing in capital investment and a good thing in margin.

Component	*Investment*	*Margin*
Raw Materials – Virgin	$$$	$$
Raw Materials – Recycled	$	$$$
Pulp Manufacturing – Virgin	$$$$$$	$$$
Pulp Manufacturing – Recycled	$	$
Paper Manufacturing	$$$$$$	$$
Converting	$$	$$$$$

Any third-grade student could immediately decide that converting is the place to be and forget the rest of the business. However, my simplistic little table here addresses the industry long term, not in the short term. In the short term, the situation can be quite variable and dynamic, and the astute investor will discern from all the conditions of the business when to invest and when to divest. This timing of investment and divestment is where money is to be made if your desk is on Wall Street.

For instance, as of the first issue of this book, early spring 2006, there was much attention being given to divestment of timber assets by companies such as International Paper[1]. With the advantages of hindsight, I personally think it was a great time to divest such assets. For now, and for the near future, timber ownership in North America, particularly timber valuable only as pulp wood, is probably a poor investment.

---

[1] This did take place throughout the industry, primarily for tax reasons, and today (2023) it is rare for an operating pulp & paper company in the United States to have significant forest holdings.

Perhaps those investing see something I don't see, such as alternative uses for growing subdivisions of cities or (longer term) in the lumber business, can find significant value, but that value does not seem to be available as solely a feedstock source for the pulp and paper industry. There are simply too many trees chasing too few digesters.

If I may digress for a moment, such opportunities have developed before—one investor suffered from myopia where another saw opportunity. This goes on all the time. A famous case is the St. Joe Paper Company in Port St. Joe, Florida. The manufacturing assets and timber assets were all held in one company and the operators of that company saw the timber assets only as future pulp wood. At the same time, they had allowed the manufacturing assets to deteriorate (more on this later). Other investors came along and split the timber from the manufacturing. The manufacturing assets, after a couple of agonizing trips through bankruptcy court, are no more. The timberland is now being marketed as the last untapped development opportunity in Florida. We'll see if this vision pans out…this particular marketing effort is getting long in the tooth (2023).

When taken as a whole, however, the pulp and paper industry cannot exist without all four components mentioned at the beginning of this chapter. Further, modern society cannot exist without the products produced from pulp and paper. Hence the industry will exist in some form for the foreseeable future. As an investor, however, you want to be discerning in your timing and targets so that you are investing in a manner maximizing your return. As a mature industry, this is often a zero-sum game, so that makes your decisions even more important and makes your personal understanding of the industry even more critical.

I hope you are beginning to see that the pulp and paper industry is not one large, monolithic, or homogenous entity. It is not a "one size fits all" enterprise. At any given moment, certain parts are thriving, and certain parts are dying. You can make money off both the winners and losers if you know what you are doing.

So, let us talk about the other three components briefly before we close this chapter.

Pulp manufacturing, that is virgin pulp manufacturing, takes trees and makes them into an intermediate cellulosic substance. This can be done in a batch process or a continuous process. The resulting products can be bleached (white) or unbleached (brown) or chemical (rayon). Please note that generally a pulp mill is designed and built to make one of the aforementioned products, not all of them. Bleached pulp can be used on site at paper machines or dried and shipped to other locations where it is "repulped" (actually just mixed with water in a gigantic blender) and used to make paper. The pulp can be shipped in relatively dry bales, rolls for diapers (fluff pulp) or relatively wet bales (wet lap). Wet lap cannot be shipped long distances or placed in long term storage as it will spoil, however, wet lap has a cost advantaged because energy was saved in avoiding removal of as much water as is removed in dry baled pulp. Water removal is costly.

Waste paper pulping is similar to repulping of baled pulp with the exception that extraordinary cleaning is required to remove dirt and contaminants, and, in the case of repulping newsprint, a deinking plant is required to separate the pulp from the ink. If you refer back to my table earlier in this chapter, you will see the pulping systems for waste paper are considerably less expensive than pulping systems for virgin wood.

Papermaking (or paperboard as the heavier grades are called—we will call it all paper here) involves taking pulp, either produced on site or elsewhere, and developing it into a product which is only 5% moisture. At the end of the paper making process, the paper will be in either roll or sheet form, depending on the grade and the next step. These very large rolls of paper (some weighing 30 tons or more) are called "reels" in most grades except tissue, where they are called "parent rolls."

Other than the tissue business, the final step in a paper mill will be either rewinding into rolls for shipment or sheeting for shipment. Due to the nature of tissue, a tissue paper mill most often has a converting plant on site where the product is transformed from parent rolls into retail packages which anyone would recognize on their local grocer's shelves. The reason tissue is converted on site is that it is a light and fluffy product and "cubes out" before it "weighs out" in transit, so it is desirable to add as much value as possible to absorb the freight costs most efficiently.

Other kinds of converting include printing, envelope manufacturing, gypsum wallboard manufacturing, box manufacturing, corrugated container manufacturing and so forth. At one time, it was popular to build some converting, such as box manufacturing, next to the paper mill. It was believed this would result in lower inventory between paper manufacturing and the box plant. However, in practice this did not happen, but what did was detrimental. Converting required more people than paper making, and these people were often of lower wage scale than papermakers. However, when both plants were joined on the same site, all workers became united under one bargaining unit and wage rates for all rose to the scale of papermaking. Since the labor-intensive part was the part with traditionally lower wages, this proved uneconomical and uncompetitive.

But as an update, one company, Pratt Industries, has reinstituted the idea of placing corrugated box plants next to the paper mills. As of this writing (January 2023) they have adopted this layout on five of their campuses and have another one under construction. One new economic consideration today is that corrugating machines have gotten so wide that the containerboard rolls that feed it are too big for standard truck shipments. Having the corrugator and the paper mill on the same site eliminates this issue for the locally made boxes.

Chapter 3
Markets

The markets in the pulp and paper industry are as many as there are grades of both pulp and paper. Some material is sold on long term contract, some is sold on the spot market. Attempts to make markets in a public way have been met with mixed success.

During the 1990's, there was a serious attempt to list and make a market in the many grades of waste paper (yes, even wastepaper has many grades) on the Chicago Board of Trade and other such exchanges around the world. These efforts, in large part, failed and for a surprising reason. For although these markets appear large, they are not as large as agricultural commodities and the traditional ways of marketing are so entrenched that a critical mass providing sufficient liquidity could not be developed. One of the reasons for this was that the big players, both buyers and sellers, remained convinced they could fair better in private transactions than in a public market. Perhaps they could and can—there is no way to know for sure.

There are published prices for most transactions in waste paper, pulp, and paper. Pulp wood also experiences a bit of this publicity, but pulp wood is absolutely a regional commodity without widespread fungible attributes, primarily due to transportation costs. The astute student will, though, recognize the falseness of these reported prices and the total lack of transparency that they provide. Namely, these prices can not be validated because they are not provided nor observable in public. Hence inquiries to sellers tend to stimulate responses on the high side and inquiries to buyers result in low responses.

Predictability of pricing is also fraught with danger. The ability to predict prices, particularly in a five year or ten-year window, is very dicey. There are some basic knowns—such as per capita consumption and the theoretical limit on tissue products and diapers (nominally 750 lbs per person per year, reached thus far only in two countries—the USA and Japan). But in specific grades, disruptions are difficult to predict precisely and can not be done without guesses. For instance, the largest pulp mills in the world are in Brazil and Indonesia. Can you tell me with any accuracy the stability of the political scene and currency in Brazil for the next twenty years? This is very important in predicting the market price of pulp. Likewise, it was predicted for years (since at least 1980) that the electronic office would kill certain printing and writing grades, but it took until the early 2000's for this to happen and then it happened quickly.

The American Forest & Paper Association publishes annual statistics on paper capacity in the USA. Their data said that paper and paperboard capacity declined 4.4% between 2000 and 2005. The survey suggested a further decline of 2.0% in 2006 and then a slight upward movement for the following two years. Such matters can be predicted so far in advance because of the long lead time to build new capacity.

The decline in capacity of the last few years was probably healthy. It eliminated some 45 + paper machines that were no doubt marginal and running to cash costs, not to rigorous P & L standards[2]. This capacity shrinkage will probably help the health of the industry in the USA for

---

[2] Many more machines in the newsprint and printing & writing grades have been shut since this book was originally published. However, many tissue and packaging machines have been built at the same time.

some time to come. The challenge is to keep shutting down old machines.

I have been saying for over a two decades now that the pulp and paper industry fifty years hence will consist of tissue, packaging, and energy production. Since I have been saying this for 20 years, we shall here and now say thirty years hence. However, since I started saying this the world has become sensitized to carbon dioxide emissions. This sensitivity casts a cloud over any energy production related to combustion of a fuel.

Be very careful of investing in the manufacture of any grades that involve using paper as a communications tool. I say this despite the known reality that very high margins are made in one specific subset of printing and writing: greeting cards. Greeting cards are the highest value combination of paper and ink on the planet. Yet, the pulp or paper manufacturers do not make these margins—it is the greeting card companies, the converters, that enjoy the profitability of this combination.

As far as the matter of integration, it means the following in the pulp and paper industry. An "integrated mill" indicates that both virgin pulp and paper are manufactured on the same site. In this way, it is not necessary to dry the pulp and ship it anywhere, so the costs of drying and shipping as an intermediate step are avoided. This usually leads to cost savings.

On the other hand, a mill with pulp making only is called a market pulp mill. A mill with paper making only is called a "non-integrated" mill. A non-integrated mill basically serves as a tolling operation, for every raw material is purchased and every output is sold on the open market.

If you remember nothing else, remember this. Paper grades whose primary function is to communicate information (in other words, something is printed on it) are dying. Paper grades used for their physical properties (boxes, packaging, tissue, kitchen roll towels and so forth) are thriving. This is the principle of using the most economically manufactured product for a particular retail function.

Chapter 4
Energy

Energy is an extremely important component to all four phases of the pulp and paper industry. Any discussion of the financial condition of the industry is incomplete without a discussion of energy.

In the raw material phase, energy manifests itself in transportation costs, whether it is from the plantation to the pulp mill or the source of discard to the recycling plant. Most of this transportation is by truck of one form or another. In the southeastern United States, the location of pulp mills is a significant factor in this transportation, and, hence, an energy issue. Many mills were built along the coast, in fact, starting at Ashland, Virginia (just north of Richmond) and extending to Palatka, Florida (just south of Jacksonville) there is a coastal mill about every sixty miles. These mills have turned out to be at a tremendous disadvantage as compared to mills further inland. Why? Their "woodbasket" (the area from which they draw raw material) is semicircular due to their coastal location. This peculiar geometry causes them to pull wood in from a further distance than if they were sitting in the middle of a forested area and could pull from a 360-degree perspective. Hence, extra costs. Now, the Aracruz mill in Brazil, formerly the largest pulp mill in the world, is also coastal and would appear to suffer from the same condition. However, since the woodbasket there grows so rapidly, this condition does not affect Aracruz.

In the pulping phase, energy consumption is usually fairly balanced, if one is pulping from raw materials (not recycled). Using a boiler called a "recovery" boiler, a virgin pulp mill can be nearly self-generating in thermal and electrical power (providing it has a high-pressure boiler with an extraction stage turbine-generator). The portion that may not be self sufficient is the lime kiln, traditionally fired with natural gas. However, today, Pet Coke (petroleum coke) has become accepted as a firing source for lime kilns, and as long as the supply/demand balance does not tip unfavorably for the pulp mills, it will provide an economical source for lime kiln firing[3].

Papermaking is an energy consumer. If it is dependent on natural gas and purchased electricity, the technical and financial matters related to energy bear close scrutiny. If it is an integrated mill, that is on the same site as a pulp mill, it can often benefit from the favorable energy balance in the pulp mill. In addition, surplus energy, from the pulp mill's perspective, can be generated with bark in a bark boiler, thus supplying the paper mill with low-cost thermal energy. All of this can, of course, be set up with an extraction turbine configuration, thus supplying electricity in an advantageous economical manner.

We are not saying much about converting and printing in this book, but it is worth noting as an aside, that the energy costs in a converting facility can also get out of hand. This is usually not so much on the electrical side as the thermal side. Many converting facilities have natural gas fired ovens to dry inks, glues and other such materials used in their processes. It is important that these ovens be natural gas fired because they are often direct fired, and cleanliness of the hot gas stream is important to product quality. In the days of high-priced natural gas, these ovens can be problematic when one is maximizing profitability or, even when trying to stay in business.

Here in 2023, all my prior thinking may soon be obsolete. The war on combustible fuels and their concomitant

---

[3] This statement was true when natural gas prices were high. In 2017, Pet Coke is barely in the picture and only at a few mills. However, it may be returning as natural gas prices are rising again (2023).

carbon dioxide emissions, may push natural gas out of the picture. The only question seems to be when.

Chapter 5
Pulp and Paper Manufacturing Assets

Pulp and Paper Manufacturing Assets come in many forms. Some people erroneously look at only the site, physical structure, location, and utilities. This is a mistake. We will look at these assets in two broad groups—tangible and intangible. It takes quality in both arenas in order to have a viable long-term business.

Tangible

By definition, objects described as "tangible" are things we can touch, feel, smell. Pulp and paper mills have plenty of these. The question always is, "How valuable are they?" This is a question that must be asked in context, and the value of the tangible assets in part relies on the intangible as well. Thus, the reader is cautioned to digest this entire chapter in order to see the whole picture.

First, I have never seen a sales book describe a pulp or paper mill as less than "modern." In fact, some chapter in most sales books will start with, "In this modern...." So how do you tell if it is modern, especially when some of the physical equipment is obviously very old?

Some people use a metric called "technological age." It is a financial formula that considers original installation date and costs, significant rebuilds including their costs and so forth. I view this as a flawed approach. Such an approach assumes that all these major expenditures, including the original investment, were done with good to excellent strategic, engineering, and financial acumen. We know this is not true, for the industry is filled with stories of poor investments, poor technological decisions and just plain dumb decisions (as of this writing, 2023, I have served as an expert witness in six lawsuits involving capital expenditure decisions, the largest of which was a suit worth $100 million [2017 dollars]).

A quick story. In the late 1980's I received an assignment to look at certain attributes of a machine located on Vancouver Island, British Columbia. I flew out there to take a look. I was engaged by the new owners of the machine, a Mexican firm. They had "purchased" the machine for free. Their cost was in taking it out and transporting it to Mexico and re-erecting it. They had x number of months to remove it from the building at which time the owners of the real estate would finish site clearing whatever they had not taken.

My assignment was to look at this machine from the standpoint of what modifications would be required to convert it from making grade "a" to grade "b." As my colleague and I poured through the records of this machine and its rebuilds, it became obvious to us, that this machine, originally erected in the 1950's, had suffered from being the last generation of everything technological. For the grades it had produced, it had been originally built to a design specification that was quickly eclipsed. Then, over the years, each rebuild had also been a last of a generation. This machine never had a chance to be profitable in its entire lifetime at that site, for it was always obsolete. And it was only built in the 1950's, a mere pup in the world of pulp and paper.

So, how do you tell the age? We'll discuss this further in a couple of chapters hence.

There are also important clues to the value of the tangible assets to the untrained eye as well. They start as you arrive at the site. Is it neat and orderly, well maintained? Are the fences in good repair? Is it a place where you would be comfortable having an office? Go to the maintenance shops—are they neat and orderly, as clean and neat and orderly as a Lexus dealership's repair bays? There is absolutely no reason they should not be.

Does the site (exterior and interior) look like someone loves it?

The ultimate visual test is to look at the water intake, or other remote structures. I was once inspecting a mill in Colombia, in the Cali Valley. I asked to see the water intake structure and building at the river. We rode there through the sugar cane fields in an armored Grand Cherokee with an army escort (twelve soldiers, four machine guns). They set up a perimeter with four machine gun nests before they would let me get out of the Jeep. But, and this is an important but—the intake structure, pump house and equipment were neat, orderly and in good working order. You can bet if maintenance is maintaining equipment under those conditions, the paper machines are well maintained, too.

Pulp and papermaking is a hot and dirty business, and I have seen investment bankers so repulsed by the overall environment that they could not discern the nuances between good and bad. If you are an investment banker, you need to practice for what you are going to see. Here's how. If you live in a large city, such a Manhattan (and you probably do) you are likely to take a subway to work. Look at the stations you come through each day. Each one of them has its own manager and those managers have their own personal spin on the housekeeping standards promulgated from on high. Start to notice the differences. Notice those that are kept clean on the main thoroughfares and those where even the corners are clean. Walk to the ends of the platforms and look up and down the tracks not in the station. You will find there an environment not unlike the basement of a papermachine building.

Another place today to become acclimated to industrial settings is YouTube. Spend some time on a site called "Virtual Rail Fan" and observe the industrial conditions there. Just like a pulp or paper mill, there a places that are loved and places that are neglected. Watch enough to be able to discern the difference.

A further word to the wise. Come to the mill dressed and equipped to inspect a paper mill. This is not Fifth Avenue; it is not even 8th Avenue. Wear jeans, a polo style shirt, and rugged shoes (steel toed shoes if you can get them). Carry a flashlight. Expect to be asked to wear a hardhat, ear plugs and full shield safety glasses (all furnished by the mill). Expect hot and humid conditions in the paper machine room, even if it is the dead of winter and you are in Grand Prairie, Alberta (I have been there when it was -40F outside and +90F inside). That's why you need a polo style shirt—you are going to want to have a coat that you can strip off as the ambient conditions change.

Often mills will have a certified appraiser go through the mill and value the equipment. Throw this away, it is worthless. Industrial appraisers are used to appraising facilities with machine tools and other such equipment that is movable. This does not apply to pulp and paper mills. These facilities (pulp and paper mills) are built into the structures and buildings. They are worth very little of their replacement cost as used machines that can be moved to another site. Yes, some items such as pumps and motors may be easily marketed, but in most cases even that is not true, because most of the old equipment on the market is so outdated from an energy efficiency standpoint that one would not want it.

Intangible

The intangible assets are often as important and sometimes (especially if they are flawed) more important than the tangible.

Intangible assets include people, suppliers, location, and most important, markets.

Asked to see the resumes of the top people in the facility—at least all the managers, and, in the technical areas, perhaps go a little deeper. From where did they graduate? Any advanced degrees? Years in the industry? Years at this site? Too much turnover is bad, but so is too little (myopic thinking develops without the injection of a healthy dose of outsiders). Ideally, turnover should be about 5% per year.

Don't ignore the hourly ranks, either. Modern mills are encouraging at least two-year degrees among the hourly ranks. Asked to see the breakdown of education of the hourly workers. If the mill does not have such records, that is a problem.

What is the mill's philosophy with suppliers—single source or multiple sources? Both philosophies can work or not work, it really depends on how they are managed. Ask for the names of several major suppliers' account managers. Call them and ask about their relationship with the mill and any insights they may have. They know more about the mill than anyone, including the management team that works there. They know all the dirty little secrets, and whether they are conscious of it or not, they have comparative opinions of the mill in which are you are interested and its peers. Now, they may be reticent to talk, based on their concern of keeping a good relationship with the mill, so you need to be a diplomat or have someone on your team with diplomatic skills make the call and ferret out the realities.

I cannot overstress the location issue. Transportation is an obvious consideration here, but there is an element that I feel is even more important. Pulp and paper professionals tend to come from rural backgrounds or at least are people inclined to an outdoor lifestyle. Hence, a pulp or paper mill in a city location may not be attractive to the best and brightest in the industry. Likewise, I have seen a location that was so rural that it was difficult to attract the best professional managers as well. This location is Prewitt, New Mexico, a site with a lunaresque terrain. Even worse, the nearest town, Grants, boasts a McDonald's, Burger King, and three state prisons, last time I counted. It offers little else—expect your connections to the contemporary world to be delivered via satellite dish and FedEx. This is where you raise your family if you are a professional in this mill. I mentioned that pulp and paper professional are outdoors types, but an important part of that equation is to have something green around. Western New Mexico along I-40 does not offer this, and hence the mill has long struggled to attract the best management team members. Conversely, it has been a great place for young professionals to "get their ticket punched." One can rise very fast in this mill and then move on. Not good for the mill, but good for the professionals...you can find them all over the North American industry.

An example of an intangible issue even a novice can spot. Several years ago, I was asked to look at some paper assets in the heartland of the US. I went to the mill and spent three days. Of course, I met all the principal players, and they knew why I was there. The overall manager of the site sat in an office on the second floor of the facility. This office was a "fish bowl," that is, it had large glass expanses on all sides. This gentleman came in precisely at 9 a.m. each morning and left at 4:30 p.m. He spent most of the morning reading the Wall Street Journal. I never saw him in one meeting, nor did I ever see him out on the operating floor. There were other reasons I recommended that my client not buy the facility, but on top of that, I told them that if they decided to buy it anyway, they needed to replace this person. They did not buy the facility.

As an update, someone, of course did buy this mill. Then, in distressed conditions in 2019, it was sold to another tissue manufacturer, reportedly for $207 million. But, and this is a big but, in the same press release announcing this transaction, it was reported that $240 million had been invested in the facility in the time frame after I told my client not to buy it (so this does not include the price paid for the facilities by the hapless investor who bought it when I said "no").

I believe this is called a fire sale.

Chapter 6
Paper Machine Width

The absolute, positively most important attribute of any paper machine.

It is a number that cannot be changed, once it is fixed. Why? A paper machine is built into the building as explained earlier. That means, as pertains to the width, foundations, beams, and beam spacing were all designed for the current machine width.

A paper machine is built to higher dimensional tolerances, than, say, for instance a bridge. If you have ever had to stop on a major bridge during traffic, you noticed one uncomfortable thing right away—the bridge bounces up and down. That is the way bridges are built and an important part of their design. A bridge is built for strength and loose dimensional stability (primarily to accommodate rapid thermal growth and shrinkage—take a mile long bridge from 90F to 35F and vice versa and you are talking feet, not inches).

A paper machine must hold dimensional tolerances, and it must do this over great widths, lengths, and temperatures. The structure of a paper machine, the foundational structure, is thus much sturdier on any comparative basis one would want to use when examining vis, a vis a bridge. Hence it (as well as all the metal components of the machine) are not easily (economically) changed once the machine is built. The Light Green Machine Institute (www.lightgreenmachine.institute) says the ratio of weight in an entire paper machine structure is 1/3 machinery, 2/3 structure and foundations.

There are even further compounding factors. If you think for a minute about the rolls and other elements spanning the width of the machine, these can be thought of as beams. The wider the span the bigger the beam, especially when it is a rotating beam (roll). The bigger the roll, the more horsepower it takes to accelerate it, maintain its speed, and decelerate it. The bigger the bearings, too. And, on and on. So, it is not economical to change the width of a machine.

So why would one want to change the width of a machine if it were feasible?

The answer is two-fold. The most important reason is to match to the contemporary width of converting or printing equipment (the second reason is economy of scale). The trimmed width of a paper machine must be integer multiples of the width of the printing equipment or converting equipment.

Newsprint is a great example. Let's say one hundred years ago, newspapers were printed one sheet wide. You went down to your friendly publisher and told him you would like to make newsprint for him. You inquired about the width of his press, got his assurance he was planning to keep it for a while, and you built a paper machine of a width to match his press. Things are going along fine. Your business is growing so you decide to go find some more. You go to surrounding towns, find the papers there are printing on the same make and model printing press as is your local publisher and business increases.

Time goes by. Printing technology hasn't changed, but you start thinking. If I built a machine twice the width of my current one, I don't need any more operators than I have now. I can make twice as much paper with the same number of operators. I am a genius! So, you build a machine that, in industry parlance, would be called, a “two out” machine, in other words, it makes reels of paper twice as wide as the market demands. Life is good, you cut your operating labor costs in half (maintenance cost stayed the same), and you start taking vacations in Florida.

Life has been good for your local publisher, too. In fact, he is having trouble keeping up with demand. While you were on vacation the printing press manufacturer visits him. He shows him new technology that allows him to make a newspaper 1/2” narrower and prints three pages wide at the same time, a huge savings for him in newsprint and printing plant efficiency and scheduling.

Your so far loyal publisher is concerned—where will he get the paper? “No problem,” says the press salesperson, “We’ve got that lined up with Mahookia Newsprint company in outer Slobovia. They can furnish exactly to the specifications of our new press.”

And you are done. Your paper no more fits the new press than anything. You are now in a “going out of business” strategy. You are stuck with the marginal customers whose newspaper business is not growing or their credit is so poor they cannot finance a new press. Obviously, eventually, you are dead—it is just a matter of how steep and slippery the slope is.

Let’s rewind the tape a bit and show another scenario. We’ll say the new press is only two sheets wide, but still a half inch narrower per sheet. If you were making a sheet that was 100 inches wide, your customer now wants one 99 inches wide. An inch doesn’t sound like much, does it? But if your operating margin was 3%, and 1 inch out of 100 is obviously 1%, your margin was just cut by 1/3. This is no better, except that perhaps you will go out of business a bit slower than the first example I gave. I know what you are thinking—I can repulp that 1 inch, so what does that hurt? Think of it another way, you just cut your efficiency by one percentage point—you are going to use the same number of operators, maintenance people, have the same depreciation, debt service, steam load and electrical load as you had before. It is a loser's game. Watch this when you are looking at buying a paper mill.

Novices get taken by this one all the time.

Chapter 7
Age and value

I have been dancing around this subject for a couple of chapters now. I'll try to bring it all together here (except for maintenance, which we will cover in the next chapter and only lightly touch on here).

The first thing that determines value in a pulp or paper mill is exactly the same thing that determines value in any other business: operating margin. It may seem simplistic and trivial, but high operating margins bring high value, or as we all know, buy low (raw materials) and sell high (finished goods).

The first complication on these statements is time (future prospects versus history or current point) and its implications for the business:

1. Is there a secure supply of raw material at a predictable and reasonable price for the duration of our plan to be in this investment?
2. Is there a secure supply of the right kinds of energy within the same parameters?
3. Are the markets into which the mill sells growing, steady, or declining? (See last paragraph, Chapter 3)
4. Is the mill strategically located in order to minimize transportation costs?
5. Is the mill located in a place that competent staff desires to live?

All of these deal with value. And by now you know that the answer to the third question involves specific grades and specific trims (paper machine widths). Every less than optimum answer to these questions indicates less value, sometimes much less value.

An example from the lodging industry points to some of the flaws that can confront the novice. In the United States, starting in the 1950's, we began building the Interstate Highway System. Prior to that there were state and federal highways, and as prosperity and mobility grew, little "mom & pop" motels had sprung up along them. The astute operators of these establishments knew that when the interstates were built, they were finished, for the traffic would be going elsewhere. Consequently, the sharp ones put their properties up for sale (before the interstate in their area was completed), pulled out the last five years' P & L's and told a great story to prospective, naïve buyers. The sophisticated sold out and left the rookies holding the bag, for as soon as the interstate was completed, their establishment was dead. Take this as an analogy. Beware of the same phenomenon if you are looking at a pulp or paper mill.

Age, aside from what I have already said about value above, is another matter.

Take the Panama Canal, for instance, or the New York Subway system. The Panama Canal is nearly 110 years old, and the New York Subway is over 120. Yet both are still the most efficient and best at doing what they do. The Panama Canal, the younger of these two, is in danger of becoming obsolete before the subway does, for there is talk of building a new canal parallel to the old one that will handle much larger ships. The day it opens, the value of the current canal falls to near zero. Update. I wrote the above paragraph in the first issue of this book. The canal upgrade happened, and the old canal is still in service. They just built new locks; the main waterway is shared by both canals.

So, age is not age. The Canal has had continuous maintenance and upgrades. A few years ago, for instance, the “bull gears” that were used to open and shut the lock gates were replaced with a modern hydraulic cylinder system—smoother to operate, less maintenance and less energy. As far as the subway system—every inch of track has been replaced several times over, as have the cars and the control system. Stations are renovated on a regular schedule. The holes in the ground may be 120 years old, but hardly anything else that is in the system dates from that time.

One must look deeper—look at maintenance and renovation projects. I mentioned poor renovation projects in the “Tangibles” subheading in Chapter 5, so unless you are very knowledgeable, being aware that renovation projects took place is not enough.

Your best bet is to hire an expert to look at the investment history and the current state of the machine(s), renovation-wise. For instance, and this is a simple one, you should not buy a linerboard machine that does not have a shoe press (or who’s price is not heavily discounted if it does not have one). I won’t bother to go into the details of what a shoe press is or the implications for having or not having one here (hint—it has nothing to do with shoes or feet), but this is just an example of the kinds of issues one better know about if they are buying pulp and paper assets.

Chapter 8
Maintenance

If the sellers use the phrases "deferred maintenance" or "breakdown maintenance" in any context other than that they are bad practices, run the other way, for you are looking at a facility with incalculable hidden land mines. (Definitions: "deferred maintenance is maintenance that is put off until some future date while "breakdown maintenance" is running until things break.)

Maintenance has become quite sophisticated over the years. First there was preventive maintenance. This is like what one does with an automobile, e.g., change the oil every 3,000 miles.

Then came predictive maintenance. This involves procedures such as mapping initial and ongoing vibration data for larger rotating machinery so that one can replace it at a time of one's choosing, not unexpectedly in the middle of the night. Another example is using a thermographic camera to take pictures of electrical connections so that hot spots can be noted, and repairs planned. A third is regular oil analysis in major lubrication systems to determine what might be going wrong and predicting failures.

Many a career has been made and many a facility sold by charlatans who assumed management or ownership roles of facilities that had been previously well maintained. These experienced shysters know they have about two years to build a P & L without maintenance and get out (for after about two years of "deferred maintenance" even the best previously well-maintained facility will start to fall apart). They look like geniuses—they are not. Whole careers have been made by sharpies jumping from assignment to assignment living off their predecessors' good maintenance practices—and leaving their successor looking like incompetent idiots.

Like everything else, though, just spending money on maintenance does not mean that good maintenance is being done. For instance, it has become a practice to turn maintenance over to quasi-permanent on-site contractors. I have struggled with this for years and have yet to see the advantage. Possibly, if a contract was properly written with bonus/penalty clauses for downtime, this would work, but I have my doubts. There just is no good substitute for a well operated maintenance program, and such a program involves close cooperation amongst and between maintenance and operations.

We will touch on this a little more in the balance sheet chapter, but one area that bears examining is the storeroom. Is it neat and orderly? Is it well lit so that one can find things? Does everything have a place and is every place well marked? Look for things with layers of dust and then compare them to the detail on the balance sheet. You may be looking at overvalued obsolete stores.

Electrical shops are a favorite of mine. Instrument and electrical technicians always have a pile of "stuff," for lack of a better word, that they are going to salvage good pieces from when they get to it. Fallacy, fallacy. If the facility has time to let a $45/hour electrician take apart old junk to find a $10 part that might be in working order, maintenance is being mismanaged. And, even if the electrician comes in and does it on his or her day off for free (highly unlikely) do you want to trust the operation of your paper machine (which probably has a gross contribution rate in the range of $25,000 per hour or more) to a salvaged part that may have been kept in less than pristine conditions from the time it was originally removed until it was salvaged? If you do, you need your head examined.

Here are some guidelines for maintenance practices. A paper machine should be shut down for a shift every three to five weeks. This small shutdown is often called a "washup" because the machine is cleaned up and some light maintenance that cannot be done while the machine is running is accomplished. Such washups need to start late on the third shift and be accomplished on day shift. It should be the practice to hold the day shift operators and maintenance over until the machine is successfully restarted (called "paper on the reel"). You are going to pay them for this, of course, but they dislike it anyway and are usually highly motivated to get back in operation. Once a year typically, there should be a "cold outage" which will last from 3 to five or more days. This needs to take place in warm weather, for the boilers will be down. Major repairs and maintenance are done during this time. More on all of this in the Chapter on the P & L.

So, in summary, watch out for:

1. Deferred maintenance
2. Breakdown maintenance
3. Contract maintenance and
4. Strange storeroom practices

All four merit very close scrutiny.

Chapter 9
Process Controls

The computer revolution has hit the pulp and paper industry just like it has the rest of society. Computer controls are ubiquitous, cheap, and necessary for a mill these days. The machinery may look old, but there is no excuse for poor computer control.

Good controls do several things. Most importantly, they monitor the raw material and additives parameters and the quality of the sheet of paper being made. On top of that, they are used for energy optimization in real time and for the purpose of bringing detailed scientific control to what was once a "black art."

Even the oldest machine may and should have nearly new control systems. The best example I can think of is a machine with some parts circa 1920, but with a control system that allows the entire machine to be monitored and controlled by an internet connection from anywhere. These folks have gone a bit far, with the idea of turning things on and off from a long distance, but the monitoring aspects are demonstrative of what can be and should be done.

Controls monitor levels, pressures, vibration, speed, energy, additives, paper physical properties and many, many other conditions. All of this data is stored in a customized data warehouse system and can be recalled easily. A "break" (unscheduled break in the paper web on the papermachine) can be analyzed easily and the root cause determined and corrected. Today's top control systems fall under the name "Industry 4.0" or IIOT (Industrial Internet of Things).

Any pulp or paper mill without modern controls is seriously out of date. One must question why this was allowed to happen and then determine if and when it can be corrected.

The other aspect of computer control is the ability to monitor logistics of incoming materials and outgoing orders. Customers today demand an "open window" to see their order status and the quality reports that go with it. Again, this is merely the entry level status of service required to obtain orders today.

Most of the controls one would recognize if they have ever been in any sort of process or chemical plant. There will be monitors (most have converted to flat panel monitors, but a few CRT's may still be around), key boards and so forth. One unique item found in the paper industry is a device called the gauging system. This is a cross machine traversing head that uses a radioactive source to discern certain physical properties of the paper. It will display and record moisture, basis weight and other pertinent data. The outputs of the gauging system will be integrated within all the other parameters monitored and controlled in the process.

As an investment banker, one needs to know these systems exist, their age, functionality and usefulness and ask probing questions if they are absent.

Chapter 10
Walk through assessments

You should, if it all possible, tour the facilities in which you are interested. You may have hired a valuation consultation (such as the author), appraisers, and others, but it still is important that you physically tour the facilities yourself. It is best to tour it with your valuation expert and then debrief this person immediately (and privately) as soon as the tour is over. After reading this far in this book, it may just be possible that you will be prepared to ask the probing question that makes or breaks your decision regarding the target assets.

And if you have read this far, you are mentally prepared for what you will see on a walk through. The more you can be prepared, the more you will be attuned to the nuances of the conditions and not overwhelmed by them on a macro scale. I hope this chapter will build your confidence in your own judgment and leave you confident that you do not have to be an expert in order to discern some basic truths about pulp and paper assets in a walk through.

You, of course, will have already reviewed the checklist found in the appendix here. You should have sent ahead the questionnaire so that the mill personnel can fill it out before you arrive.

Overall housekeeping is the first key to the status of the facility. It may have not been painted recently, but there is no excuse for it not being neat and orderly.

Look for differences in the accommodations for office personnel and hourly workers. The differences here should be as to functionality and not have a hint of status. In other words, restrooms or locker rooms used by all personnel should be of the same level of housekeeping, neatness, and repair. Discernable differences are potential clues to labor problems.

But, I have jumped ahead of myself. Start with the parking lot. Does the mill have reserved spaces, and if so for whom? There should be reserved spaces near the front door for guests (of any type—customers, suppliers and so forth). Also, there may be reserved spaces for winners of safety awards or other such incentives. And, of course, there will be government mandated handicap parking spaces. However, if there is an excess of spaces reserved for salaried people of title, watch out. Some things valued in this mill are not necessarily related to contribution to the overall effort. The ideal is to have no reserved spaces.

If your path into the mill does not take you through the guard shack, ask to see it anyway (there are more than likely more than one—look at all of them, especially the most remote). Look the guard shack over. Is it neat and orderly, everything in its place? Does it appear that emergency procedures are prominently posted and prepared for immediate implementation? All are clues to the competence of the overall management of the mill.

None of the balance of the matters here are in particular order. In fact, they should just be observed as you allow yourself to be guided through the facility. It is better that you do not tip your hand as to what interests you, so that your guides start trying to out think you. You want them to be relaxed, free and easy, so that they will tell you the truth, not what they think you want to hear.

What you are looking for is hints as to the management's attitude about running the facility. The finesse on this is to determine if this attitude is local or from on high. This will make a difference to you depending on whether you are buying an entire company or just one asset. For instance, many a facility manager wants to do things in a very professional way yet does not have the budget to do so. In other cases, a facility has had a string of poor managers and corporate headquarters has decided to cut themselves loose from what they perceive as a hopeless situation. Your own situation is unique, and you need to assess these given conditions and how they will play out with your objectives should you make an acquisition.

Next, how is the overall facility laid out? Were the original designers and their successors logical in their approach? Do materials seem to flow logically from one end of the facility to the other? You will not be able to do anything about these conditions, but they may be a competitive issue.

Notice the location of maintenance shops and store rooms. Are these of a "satellite" fashion, i.e., a small maintenance shop near each primary production area? Is there a central shop? Does it appear to you that there is an extraordinary amount of maintenance traffic? Do you observe maintenance excessively using freight elevators to move heavy materials between operating floors? Maintenance is a mill's largest labor force and how they, and the things they need, are organized, transported, and managed is of paramount importance.

Look at laboratories. There may be several throughout the facility. In some cases, they may be for environmental matters, in others for paper quality matters. They may be separate or together. The paper testing laboratories should be at "TAPPI Standard Conditions." This means they are kept at a prescribed temperature and humidity. Entering such laboratories is through an air lock. Somewhere in the laboratory should be instrumentation recording the temperature and humidity conditions. These precautions are necessary to test paper properly. Since paper absorbs moisture easily, you can see why it is necessary to do all testing at specific conditions for otherwise the results are meaningless. Look around each laboratory. Is it neat and orderly? Does the instrumentation look modern, or does it look like it came from a 1950's Frankenstein movie? If modern, are the instruments interconnected on a LAN (Local Area Network)? How is the data recorded and stored? Ask to look at the path of data as recorded through to delivery to the customer. If there is any confusion or a tortuous path here, you have just seen a possible hint of management problems.

Chapter 11
P & L reveals its secrets

The typical pulp and paper mill Profit and Loss Statement will contain the following categories:

P & L

Operating Hours Available
Operating Hours Scheduled
Scheduled Downtime (Hours)
Operating Days Available
Operating Days Scheduled
Tons Produced Per Scheduled Day
Cull/Trim Per Scheduled Day
Tons Sold Per Scheduled Day
Total Tons Produced
Inventory Change
Cull/Trim Loss
Total Tons Sold

Gross Sales
Commissions
R & A's
Transportation
Net Sales

Pulp/Waste Paper

Variable Costs

Chemicals
Electricity
Steam Fuel
Machine Clothing
Maintenance
Hourly Labor

Hourly Fringes
Inventory Change
Other
Total Variable Cost of Sales

Fixed Costs
Salaries
Salaried Fringes
Taxes and Insurance
Management Fees
Other
Total Fixed Cost

The above is from an actual paper mill P & L. Let us tear this apart, line by line, and see what mysteries we can reveal.

Operating Hours Available
Operating Hours Scheduled

Let the games begin. What the raw numbers are here is not nearly as important as what management tells you about them. Typically, management will brag about "efficiency" or "percent uptime." These are meaningless phrases unless you know what the standard against which they are measuring is. The typical mill is scheduled to run 360 days per year (some use 355, others use numbers as high as 362). It is important for you to know what this number is and to understand, generally, that that number is 360 days. Now you can do the math: 360 days are 8,640 hours. So, on an annual P & L, if the "Operating Hours Available" deviates much from 8,640 hours, ask questions.

What happens to the other 4 or 5 days? Those are the annual outage.

Typically, the Operating Hours Available deducts the annual outage from the full year's calendar (either 365 days or 366 days if a leap year). The wash ups, mentioned earlier, are taken out of the "Operating Hours Available" resulting in the "Operating Hours Scheduled."

The annual P & L accounting for hours is one thing. The monthly, of course, depends on how many days are in the month, how many wash ups, and if the annual outage occurs that month.

Scheduled Downtime (Hours)

This should mathematically follow from the arguments above. If it takes more than two sentences to explain, someone is blowing smoke.

Operating Days Available
Operating Days Scheduled

On a yearly basis, these are 1/24 of the Operating Hours Available and Scheduled. On a monthly or weekly basis, they are calendar dependent and dependent on the actual occurrence of scheduled downtime. However, all weeks and all months for the year must add up to the total for the year.

Two special, tiny "watchouts" are the months and within those months, the weeks that time changes occur due to daylight savings time change over. Actual hours in those weeks are skewed by one hour against the calendar.

Tons Produced Per Scheduled Day

Sometimes this is called "scanner tons" or "gauge tons" which means the number of tons the radioactive gauge (see Chapter 9, Process Controls) says was produced. No matter what it is called, it should be the gross number of tons produced divided by the number of scheduled days. You also need to know if the mill is producing "short tons" (US English Measurements) or "long tons" (Metric Measurements) as the difference is 10% and vital in measuring from mill to mill of like grades.

Cull/Trim Per Scheduled Day

"Cull" indicates off specification. This will also be measured in tons. Trim is tons that may be of standard specification but is in a "trim" or "side" roll. This occurs when the orders do not "trim out" the machine (See Chapter 6, Paper Machine Width for more on this).

Ask to see the detail on Cull vs. Trim. If Trim is high (over 15%) you are clearly dealing with an obsolete machine for the grades produced. If it is between 5% and 15% it is worrisome and merits a detailed discussion.

Tons Sold Per Scheduled Day

Of all the tons produced, good tons and cull/trim tons, these are the tons for which the mill has orders (some cull/trim tons are sold at a reduced price, some are taken back to the beginning of the paper machine and repulped). The mill may be making to inventory, so, other than not exceeding the tons produced per schedule day, it may or may not have anything to do with the previous numbers. You must dig into these numbers to know what is real.

Total Tons Produced

Hopefully you are not lost and have not given up on this exercise! You can see how confusing it can be. Now, "Total Tons Produced" are those tons, good, cull, trim, sold and unsold that were manufactured and ended up in the warehouse (in other words, were not repulped).

Inventory Change

All the while the paper machine is producing product, trucks and railcars are loading out of the warehouse. If, in a given period, more product was produced than shipped, then inventory went up. If less product was produced than left, inventory went down. Depending on local practice, negative or positive inventory swings mean different things—you must know whether locally a negative number is an inventory increase or decrease. The only way to know is to inquire.

A special caution: some mills have inventory stored off site. You may not think about it because you can't see it. You need to know what it is and whether it is counted as sold or not. Plenty of games are played with inventory you can not see.

Cull/Trim Loss

This is all the material that is repulped (not sold at any price) within the period of interest. It may be expressed in tons or as a %.

Total Tons Sold

These are all the tons with an owner other than the mill that have been produced in the time frame covered by the P & L. They include good tons, cull tons and trim tons with an owner other than the mill. This does not include tons made and put into inventory that are not yet sold. Why would the mill put tons into inventory that are not yet sold? There are several reasons. One is that in their normal order run, it is the most efficient time to make those tons. Secondly, their order book may be light. Thirdly, they may have a customer which buys regularly for which they stock tonnage but do not record as a sale until they get a specific purchase order. These reasons can be legitimate, but all of them can be abused by a mill that is in trouble. Dig deep for the truth.

Gross Sales

We finally get to currency. This number will be, hopefully, both in dollars and in dollars per ton for the period of interest. It is the gross amount received.

Commissions

For sales made via brokers, and even sometimes those of internal sales people, depending on their compensation package, commissions may be paid. These are a deduct.

R & As

This means "Returns and Allowances." You may think you have sold the paper when it leaves for the customer, but it does not stay sold until the customer says so. Sometimes the quality is poor, and they return it. Sometimes (bad practice) the mill gives them a discount for working with off spec paper. This is a bad practice because customers can easily get in the habit of seeking a discount after the sale. Preferable treatment is to tell them we don't want you to use off spec product—send it back. Guidelines here: R & As should be less than 2% of Gross Sales.

Transportation

Paper is sold F.O.B. purchaser, so the mill pays the freight. This means a higher priced order may be less valuable to the mill than a lower priced one if the order delivery point is further away.

Net Sales

Sometimes called "mill nets." This is the price per ton of all sold paper after all deductions. Nothing really counts until one gets here.

Pulp/Waste Paper

This is what one pays for the primary raw material to the paper machine.

A couple of cautions:

1. If this paper machine is on an "integrated" site, i.e., a site where pulp is produced from wood chips, this number is a phony baloney transfer price. It is meaningless unless one looks at the overall P & L for the entire site.
2. If this pulp is purchased pulp, it is purchased in metric tons (or "tonnes") no matter what measurement system the mill uses to measure its own production. Get this correct…there is a 10% difference in weight of metric versus US tons.
3. If we are talking about purchased waster paper, it may or may not be purchased in metric tons (depending where on the planet the mill is), but there is another consideration. Is this the price before pulping, as received? There will be losses of 5 to 20% depending on the type of waste paper and the grades being manufactured on site. These can not be avoided.

If you thought there were plenty of fun and games thus far, now it really gets interesting as we look at variable costs.

Chemicals

Chemicals, unless the mill manufacturers a coated grade, are fairly innocuous and uninteresting. These will be chemicals to help with the "water health" of the machine, boiler chemicals, environmental plant chemicals and so forth. Minor stuff, but good chemical sales people can load a mill up with all sorts of nearly useless chemicals that are very expensive.

Electricity

Electricity can come from two sources and often from two or three sources in the same mill. There can be purchased power and there can be power generated on site via turbines and via solar panels. Paper mills are popular places for self generation, for certain attributes of steam properties can be economically applied in a paper mill.

Items of concern here are long term power contracts and the peculiarities of these. If electrical power is a large cost, consider engaging a power contract specialist to review any contracts.

Note: this may not necessarily be a variable cost. If the mill has been experiencing extraordinarily poor running conditions, this behaves more like a fixed cost. For the motors and so forth do not know if the mill is making good paper or not.

Steam Fuel

This can be of several sources or only one. Steam fuel can come from:

Chemical pulping (recovery boiler)
Bark
Fuel Oil
Coal
Natural Gas
TDF (Tire Derived Fuel) and/or

anything that will burn.

This is, of course, not all the costs of operating the boilers, but only the fuel costs.

Machine Clothing

The term describes the wires and felts which are the surfaces that touch the fiber in the wet end of the paper machine. These wear out on a regular basis and must be replaced. There is a whole field of science associated with these components. But like with electricity, see this note:

Note: These may not necessarily be a variable cost. If the mill has been experiencing extraordinarily poor running conditions, these behave more like a fixed cost. For the machine clothing does not know if the mill is making good paper or not.

Maintenance

An area of extraordinary land mines. Some major maintenance may be capitalized, so this category must be examined in conjunction with capital expenditures. Some items in stores may be on consignment and hit the books only when used. So, just because you saw it on your tour does not mean it is an asset of the facility. And there will no doubt be things in stores that are carried at a much higher price than they are worth—they may be obsolete, worn out, or just plain old. The worst case I have see is an overstatement of about $4 mm in a mill worth about $20 mm in total. Close examination is merited if you are considering buying. Also, most maintenance is not variable—

it is fixed—so previous notes of caution apply here as well.

Hourly Labor
Hourly Fringes

Not much to mess with here—what you see is what you get unless an extraordinary amount of contract labor is being used and hidden somewhere else.

Inventory Change

An obvious catch all. We already covered the important inventory change issues earlier.

Other

When we can't find any other place to bury it.

Total Variable Cost of Sales

Well, it is, and it isn't—remember my cautions in the categories above.

Fixed Costs

Salaries
Salaried Fringes
Taxes and Insurance
Management Fees
Other

Total Fixed Costs

These are all usually fairly straight forward, hence you will not find much of interest here, except maybe salesperson's cars, home offices, administrative help and so forth that is not on sight and hence out of mind.

A general caution—look for things not on site and where they might be buried in the financials. These include warehouses, sales people, the boss's girlfriends' apartments, and so forth.

Chapter 12
What is on the balance sheet?

We will not go through the agony that we did with the P & L to discuss the balance sheet. There are just a few places of concern here, and they are relatively straight forward.

First, we have the issue of overvalued assets. Pulp and paper mills, being large assets with continual improvements, tend to always have a hefty value on the balance sheet, even if the overall facility is old. Owners are loathed to write these down to current value. Sometimes perhaps GAAP accounting may even prohibit them from doing so. However, that does not mean for your purposes you cannot adjust for you own analysis.

Examples? Papermachines serving declining grades. I don't care if it was installed brand new three years ago; for instance, I doubt that a new newsprint machine will ever earn a decent return unless located in a very special place on this earth. It is probably on the books for far more than it is worth. Update, 2023: there are not many newsprint mills left, so not so much to worry about today.

We already mentioned (maintenance) stores inventories in the last chapter. *I have never been in a mill that did not over value its stores inventory.*

The same thing can happen with finished goods inventory. You should not only demand an independent hand count, but you should also look at the age of that inventory. I once walked in a mill, which at the time was bringing in gross revenue of $120 mm per year. The finished goods inventory claimed on the books was $8 mm. Not only was it old, about half of it did not even exist.

Another interesting phenomenon one may see is a machine sold and leased back. It became popular in the 1990's to sell paper machines within mills in the United States to European companies and then lease them back. The reason for this was that the Europeans could depreciate them faster in their home countries. The new buyer and the new leasor then split the tax savings caused by the rapid depreciation.

On the liability side, you can be almost certain that environmental clean up of the site, should it be closed, has not been recognized as a liability. If the site is over twenty-five years old, even the removal of asbestos insulation has probably not been fully recognized. If you are told the asbestos insulation has all been removed from the site, asked to see independent proof certifying this.

Contracts may run for some length of time. These may be contracts for energy, raw materials and so forth.

Likewise, future Returns & Allowances (see previous chapter) are most likely not on the books.

All of this, of course, leads to one conclusion: buy the assets, not the business. Yet, even in this case, when it comes to the environmental issues, the new owners will not be able to escape all the environmental liabilities.

Chapter 13
When to buy, when to sell

The trivial answer, already mentioned earlier, is buy low and sell high, but the trick is how do you know? Hopefully, by now you have learned a little about the pitfalls of this industry as well as the opportunities it presents.

Buy, of course, when the current owner is in distress or has chosen to take a different strategic direction. However, do so with caution.

Do not buy assets in strategically declining markets, such as newsprint. A good rule of thumb is to be concerned about buying into any market where the primary end use of the paper is communications of thoughts and ideas, i.e., any paper used for printing of either images or text.

Some argue that packaging is in decline, but, at least as of this writing, I do not think so. Corrugated containers (the brown boxes with the wavy inner layer) are often said to follow consumer goods. The logic goes that consumer goods manufacturing has moved out of the country (the United States in this case), hence that is where the boxes are needed. There are at least two counter arguments to that: (1) 25% of all manufactured goods in the world are still made in the United States as of 2006 and (2) Internet purchases have driven heavier use of smaller boxes for shipping (such as one or two books at a time from Amazon).

A couple of changes to be noted in 2023 are these. Manufacturing is moving back to the United States; hence box manufacturing is growing here. Amazon is using more bags and pouches, but their business is still growing overall. The Covid pandemic drove more people to home delivery, requiring packaging of all sorts.

So, first and foremost, buy assets in grades that are stable or growing.

Next, look for strategic sales. Some firms may be in a number of grades of paper and have decided that certain ones no longer meet their strategic vision. The assets associated with these grades sometimes sell cheaply.

A number of years ago, a major forest products company decided to get out of pulp and paper. The obviously good assets sold quickly. One ugly duckling was left and sold at a discount. That ugly duckling has turned into a beautiful swan, able to make a multiple number of grades very well and without compromise as compared to their competitors in each grade.

Another source of good buys is the small player that has gotten into cash flow trouble. Not being able to maintain sufficient credit lines, they are in desperate straights.

The source of these desperate straits is, of course, of great interest. Perhaps they have a poor energy situation, and you see a way out of that. Perhaps it is a labor contract that has put them in trouble. A new owner may have a chance to fix these matters.

One situation to never buy into is this. A large company decides to exit a certain grade, and absent credible buyers, decides to encourage an employee buyout. These almost always are bad deals. The reason the current owner cannot find a buyer is that the facility is a strategic loser because of grades, asset conditions or other serious issues. The reason the employees want to buy it is to save their *all* of their high paying jobs. These almost always fail and are almost always a bad bet if you contemplate becoming entangled with them.

It is of value to understand how decisions are made to build pulp and paper assets and what happens to markets and values as they are built. Consider this diagram:

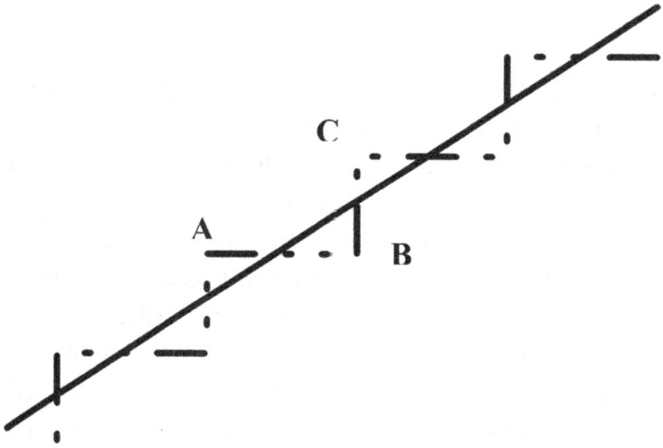

The steadily increasing sloped line (the solid line) represents the gradual increase in demand of a specific grade of paper; let's say toilet tissue (a grade whose increase in demand is directly related to population growth). The "stair step" dashed lines represent capacity. Everyone in the business knows the demand and capacity—these numbers are quite transparent to the entire industry.

For purposes of this discussion, we will enter at point "**A.**" At point **A,** the industry has just completed a major round of expansions. The capacity line is above the sloped demand line and prices have fallen, or at least are not rising at a rate equal to the immediately previous period. As time goes on, due to unattractive pricing, capacity remains flat, as is evidenced by the flat capacity line between **A** and **B.** Prices, however, have not remained constant, in fact as soon as line **A B** crosses the demand line (what happened was that demand grew to absorb the new capacity) prices start rising rapidly. Strategic planning departments at all the tissue companies wake up. All order new machines and install them at point **B**. Capacity shoots to point **C**, the wrong side of the demand curve, and prices are depressed. The cycle repeats again and again.

If you are not familiar with the pulp and paper industry, you may think I am stretching your credulity to suggest that this phenomenon happens as lockstep as I have shown it. Let me assure you, it does. And, at the same time, it is so lockstep that it puts the second derivative players not shown here (the equipment manufacturers) through endless turmoil. Their shops are either full or they are empty—there is almost no in between. The first player to order a papermachine in the cycle gets a bargain (because the workshops are empty), all the rest pay retail. It has happened over and over again.

Early in this book, I mentioned plastics and chemical manufacturers and stated that they are different from the rock, paper, and scissors crowd. This is true, even though they follow a similar stair step/demand slope diagram. The difference is the scale of the size of their plants is much smaller and hence the capacity increments are much smaller. This allows them to hug the demand line much closer, resulting in less disruption than experienced in the rock, paper, and scissors industries.

When to sell? That's simple—when you are absolutely convinced the business will be worth no more tomorrow or any other day in the future than it is today. Now read the previous sentence carefully, for it has nothing to do with the price at which you bought, for you may have made a bad deal and bought at a price much higher than you should have. However, this little book has hopefully helped you avoid that condition.

Appendix I

Checklist

This checklist is designed to prepare for and conduct a walk-through of an acquisition target. It should prove valuable to both buyers and sellers, for if a buyer asks for this information, the astute seller will be ready to provide it.

Name of Facility

Location

Proposed date of visit

Pre-visit request list:

1. P & Ls for past twelve months
2. If integrated (both pulp and paper manufacturing on the same site), what is the mechanism for setting transfer prices between the pulp mill and paper machine(s)?
3. Customer list (ask for mill contacts' names and phone numbers)
4. Returns & Allowances detailed history for the last five years
5. Trim roll/side roll history for the last five years
6. Large supplier list (ask for account managers' names and phone numbers)
7. Utility contracts and bills
8. Site plan
9. Process plan or flowsheet
10. Executed capital plans for the last ten years (audited, if possible) and projected for the next five.

11. Organization chart, with everyone on it
12. Resumes of all managers and technical professionals (may be difficult, but try to get them anyway)

During the visit:

Weather conditions:

1. Temperature
2. Wind
3. Precipitation

Appearance from public thoroughfares

Proximity to nearest neighbors

Nature of neighborhood (industrial, residential, rural, river, lake, ocean)

Does any condition in the neighborhood appear to be a future liability?

Condition of:

1. Parking lots
2. Administrative offices
    a. Are administrative offices far away from the production operations?
    b. Is there a sense that the occupants of the administrative offices do not go into the operations?
3. Restrooms
    a. Is there a differentiable appearance or state of condition of administrative vs hourly restrooms?

    b. Would you use the restrooms you have seen on this tour?
4. Each operating area
5. Maintenance shops
6. Storerooms
7. Things stored outside
    a. Are they neat and orderly?
    b. Are the ones not normally visible in a poorer state of organization than the ones that are?

What are the preventive and predictive maintenance programs? Are they formal?

How do laboratories look? Are they to TAPPI standard conditions?

Review data collection paths from the machine through the labs to the customer.

If you have read this book and observe something you think is unusual, it probably is. Ask about it!

Websites operated by Paperitalo Publications, LLC can be found at www.paperitalo.com

Made in the USA
Monee, IL
04 November 2026

67384078R00046